Where On Earth Is Mama?!

By Mutasem Moussa

Where On Earth Is Mama?!

By Mutasem Moussa

Illustrated by Sheri Meshal

Acknowledgement

I am Mutasem Moussa. I wrote this book when I was six, but I am currently 11 years old. I live in a small town in Illinois called Hickory Hills. I would like to thank my teachers for loving my book. I would like to thank my aunt for illustrating this book, and my friends and family. This all started when I was wandering around on the Internet and found this crazy idea about a space elevator. I was like, "I would like to write a book on this topic!" Then my aunt and I went to work on the names, museum and the other ideas. I would like to thank my brother Baraa for being an amazing brother. Also, my sister Noor for being loving and really funny for what she says. I would like to thank my mom and dad for always loving my creations. Finally, I would like to thank my Aunt Sheri for creating my pictures and loving my creations like my mom and dad. As I said, I am only 11 years old, and I hope this book will be nice for your children or yourself.

Once upon a time, there was a boy named Max who wanted to visit the Blast Off Space Museum more than anything in the whole world.

On his seventh birthday his mom put on her red coat and told him, "Get in the car, Max. I have a surprise!"

"Yippee! Where are we going?" Max shouted. He knew it was someplace cool, because his mom was so cool she let him cut his own hair.

"It's a surprise. Put this blindfold over your eyes," His mom said, handing him a red handkerchief.

"Okay, Mama! You're the boss."

"Okay Max, you can take off your blindfold."

Max took off the handkerchief and unfastened his seatbelt. He blinked, but couldn't believe his eyes. "Mama, it's the Blast Off Space Museum!"

They locked the car and ran inside. He went to see the sun first through a giant telescope. It looked like a huge fire exploding. It was very bright.

Max couldn't see very well when he was done. He couldn't see his mom. There were two thousand people there. Max was scared when he couldn't find his mom. Then he saw someone in a red coat up on the second floor and he knew he had to get up there fast.

Max stumbled along until he found an elevator. He couldn't read the buttons inside, because his eyes were still seeing stars from looking at the sun. The doors closed and the elevator moved.

The elevator moved so fast that he fell down. It went up and up and up. It took forever for the elevator to stop. Max fell asleep it took so long. When he woke up he looked at the button he'd pressed and freaked-out when he read the word 'Space' just before the doors opened.

Max gasped when he saw the moon up close through a giant window. He stepped out of the elevator. He was in a space station! He was hungry after the long ride and spotted a machine that had all kinds of astronaut food. He pulled out the money his mom had given him and bought a chocolate space bar.

A worker saw him and said, "Hi, you must be lost." "Yes!" answered Max.

The worker told him to go to the second floor. "Just press Second Floor Museum. There's a lady looking for her son in the gift shop. She's very worried. That must be your mama."

Max told the man, "Thank you! I have to get there right away!"

Max went back inside the elevator and straight to the museum gift shop, where he found his mom was buying a leash for him.

Max's mom said, "Max! I've got another surprise for your birthday!" She snapped the leash to the zipper on his coat and tugged on it.

"Mama! Max shouted.

"Happy Birthday Max!" his mom said and gave Max a big hug.